A Space Story

First Edition

A Space Story

by
Karla Kuskin

illustrated by
Marc Simont

Harper & Row, Publishers
New York, Hagerstown, San Francisco, London

Before he goes to bed every night at 8 o'clock Sam has a glass of water. After that he says that he is too hot and then he asks his mother as many questions as she will answer.

"How long does it take to walk around the world?"

"What color is rain?"

"Do people eat camels?"

Tonight Sam is looking out the window at the night sky.

"How many stars are there in the sky?" he asks his mother.

"Billions and billions," she says.

"Do people live on them?" he asks his mother.

"People can't live on stars," she says. "But I suppose there might be some people out there on planets somewhere."

"I wonder what they're like," Sam says, but his mother just says, "Good night my boy," and softly shuts the door.

Outside Sam's bedroom window the lawn is too dark to see the dandelions. Beyond the dark lawn is the dark garage and across the dark street is the empty lot with three trees. All dark. Above the dark tops of the trees a few dark birds fly to their warm dark nests. Above the dark birds dark airplanes blink bright red, green, and white lights. Above the airplanes the white moon lights the clouds, and far away above the moon the arc of the sky stretches farther than you or I can possibly imagine. It is filled with planets and moons and stars without number.

Ninety-three million miles from Sam's bedroom window is the sun. The sun is a fiery star. On a very hot day in New York City the temperature is about 95 degrees. At the center of the sun it is 15,000,000 degrees. The sun is the star closest to the Earth.

Tonight it shines brightly on the other side of the world. Tomorrow it will rise and shine on Sam's side of the Earth. The sun is made up of bright hot gases. Nothing and nobody could live on it. It is always on fire.

High in the sky beyond Sam's bedroom window on the planet Earth, there are 8 other planets turning around the sun.

Pluto
Saturn
Uranus
Neptune

Mercury is the smallest one. It is also the one closest to the sun. During the day it is hotter than the hottest oven. There is no air on Mercury. No one like Sam could live there.

High in the sky beyond Sam's bedroom window is Venus. As Venus turns around the sun, it is the planet that comes closest to Earth. At night it shines more brightly than any of the other planets. Venus is surrounded by clouds, so we can only guess what it looks like. We do know it is very hot and probably covered with lava. The air is thick and poisonous. No one like Sam could live there.

High in the sky, on the other side of the Earth, there is a planet that looks red from Sam's bedroom window. It is Mars. Mars has a mountain higher and a canyon deeper than any mountain or canyon on Earth. Maybe, very long ago, there was water on Mars, but now it is too cold and there is not enough air for anyone like Sam to live there.

High in the sky beyond Sam's bedroom window on Earth, even farther than Mars, is the planet Jupiter. It is the largest of the planets going around our sun, and it has 12 moons. On Jupiter there is nothing solid to sit on and nothing solid to stand on. No one like Sam could sit there or stand there or live there.

High in the sky beyond Sam's bedroom window, out in the night beyond Jupiter, is the planet Saturn. The 4 shining rings around Saturn are probably made of rock and ice particles, but because they are so far away we cannot be sure. Saturn, like Jupiter, is not made of solid stuff. No one like Sam could live there.

And far, far, far into the sky beyond Saturn 3 more distant planets turn around the sun. Of the 3, the one closest to Earth is Uranus. It is extremely cold and there is no air to breathe, just gas. Neptune is even farther away and colder. It would take 9 years for a space ship to get to Neptune, but what would the people breathe once they got there? There is no air to breathe on Neptune either. The last planet in our solar system is Pluto. It is so far from us, so dim, so cold, that all we know about it is that no one like Sam could live there.

Beyond Pluto the stars stretch out. Little and big, bright and dim. Stars without number. And then beyond those stars there is still more space—more moons, stars, planets, galaxies. And somewhere way out in the distant spaces of the sky, a faraway planet spins around a faraway sun.

There is a dark street on that faraway planet and an empty lot with three trees. All dark. Across the street the lawn is too dark to see the dandelions. Next to the dark garage there is a dark house. All of this is far, far away.

And far, far away a boy looks out the window at the night sky.

"How many stars are there in the sky?" he asks his mother.

"Billions and billions," she says.

"Do people live on them?" he asks his mother.

"People can't live on stars," she says. "But I suppose there might be some people out there on planets somewhere."

"I wonder what they're like," the faraway boy says, but his mother just says, "Good night my boy," and softly shuts the door . . . far away.